# MARCH OF THE PENGUINS

Translation Copyright © 2006 National Geographic Society

Translated from *La Marche de L'Empereur* by Luc Jacquet with photographs by Jérôme Maison (published in France in 2005 by Éditions Michel Lafon) and adapted from the narration of *March of the Penguins* by Jordan Roberts

Translated and adapted by Donnali Fifield

Photographs Copyright © 2004 Jérôme Maison/Bonne Pioche
Except p. 144 bottom, p. 145 center, p. 147 upper left, p. 151 top, p. 155 center, p. 157 upper left and bottom, p. 158 center, Copyright © 2004 Laurent Chalet/ Bonne Pioche

Published by arrangement with Éditions Michel Lafon, and Bonne Pioche/APC and Warner Bros. Entertainment Inc.

Design and production for Édition Michel Lafon's *La Marche de L'Empereur*: Vandepub by Mathieu Thauvin; Editorial direction by Marie Dreyfuss, with the collaboration of Judith Carraz

WARNER INDEPENDENT PICTURES AND NATIONAL GEOGRAPHIC FEATURE FILMS PRESENT A BONNE PIOCHE PRODUCTION IN ASSOCIATION WITH WILD BUNCH 'MARCH OF THE PENGUINS' AS TOLD BY MORGAN FREEMAN MUSIC BY ALEX WURMAN EDITOR SABINE EMILIANI CINEMATOGRAPHY LAURENT CHALET JÉRÔME MAISON EXECUTIVE PRODUCER ILANN GIRARD PRODUCED BY YVES DARONDEAU CHRISTOPHE LIOUD EMMANUEL PRIOU BASED UPON THE STORY BY LUC JACQUET BASED UPON THE SCREENPLAY BY LUC JACQUET & MICHEL FESSLER NARRATION WRITTEN BY JORDAN ROBERTS DIRECTED BY LUC JACQUET WITH THE PARTICIPATION OF BUENA VISTA INTERNATIONAL FILM PRODUCTION (FRANCE) AND CANAL + A CO-PRODUCTION WITH APC IN ASSOCIATION WITH THE FRENCH POLAR INSTITUTE (IPEV)

BONNE PIOCHE    NATIONAL GEOGRAPHIC Feature Films    www.marchofthepenguins.com    G GENERAL AUDIENCES All Ages Admitted    WARNER INDEPENDENT PICTURES © 2005 Warner Bros. Ent. Inc.

Library of Congress Cataloging-in-Publication Data available on request.

ISBN : 0-7922-6182-8

Printed in U.S.A.

# MARCH OF THE PENGUINS

Luc Jacquet

including narration written by Jordan Roberts

photographs by Jérôme Maison

Translated and adapted by Donnali Fifield

NATIONAL GEOGRAPHIC

WASHINGTON, D.C.

You need only watch an emperor penguin in the water to be touched by the strange fate that nature has reserved for him. In the ocean, everything is easy for him; he is powerful

# FOREWORD

and sleek. With a smooth leap, he springs up from the water onto the ice and is suddenly transformed into an awkward waddler. Why can't this child of the frozen oceans stay in the water, where he is clearly so much more at home? That is the emperor penguins' singular predicament.

Like all species, the emperor penguins must ensure the survival of their kind; they must lay an egg in a dry place, incubate the egg, and raise the chick that hatches from it. But in the Antarctic, few places are favorable for reproduction. The emperor penguins therefore have no other choice but to bring their young into the world on the ice, where no other creature dares challenge them for this dubious privilege.

In the winter, the continent of Antarctica is surrounded by ice, a thick layer of frozen sea up to 70 miles (113 km) wide. This block of ice looks immovable, but it isn't reliable; storms can cause it to shift at any moment. So for the nine months necessary for reproduction, the emperor penguins must settle in areas far from the sea, much farther to the south, where the ice is stable.

This is what leads to the strange fate of the emperor penguins: They must live divided between the ocean, which feeds them, and the ice, which is a barren but safe desert, a true oasis for their chicks. The emperor penguins march back and forth between the two destinations, trudging hundreds of miles despite the winter and the dangers, marching between the nourishing sea and their babies, who are hungry. Each year, the emperor penguins have a rendezvous with winter, to begin again what their fathers and their fathers' fathers have done before them.

*March of the Penguins* tells the story of this epic journey.

LUC JACQUET

The average temperature, here at the bottom of the Earth, is a balmy 58 degrees below zero(-14.4°C)—when the sun is out. It wasn't always like this. Millions of years ago,

# LEGEND HAS IT

Antarctica was a tropical place, densely forested and teeming with life. But then the continent started to drift south. And by the time it had stopped, the dense forests had all been replaced with a new ground cover: ice.

All the former inhabitants died or moved on, long ago. Well, almost all of them.

Legend has it that one tribe stayed behind: the emperor penguins. These stalwart souls refused to leave. For ages, they have made their home here—the darkest, driest, windiest, and coldest continent on Earth. Children of winter, they are the only creatures able to take on the unleashed elements of their homeland. Each year they must march. To give up is to die; to trudge on is to survive and bear their young. Stubborn, heroic, they prevail over the cold, the isolation, the icy vastness.

Here, the blizzard blows—its moan echoes far beyond the horizon, which seems improbably remote. The snow falls, monotonous. The flakes, like confetti, are instantly caught by the powerful wind. In the air drifts an odor of steel. The ice cracks, splits, the fissures making a short, hard click as they break apart.

In this frozen landscape the emperor penguins must march. Each year, foolhardy, they take on the challenge of the ice. So this is a story of survival, a tale of life over death. But it's more than that. This is a story about love.

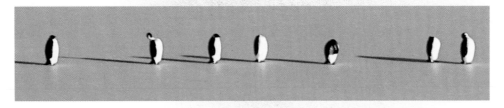

*In March, when the ice begins to form, the emperor penguins gather together and*
*leave the open sea. It is the beginning of their long march across the white desert.*
*Where they are going, no other creature could survive the winter.*

Anyone who really wants to understand this most human of birds must first love them. The emperor penguins are at home in the ocean. Prodigal sons of the sea, they are graceful in the water. They play and show off; even in bad weather, they perform the most creative of ballets. But at the start of each winter they have to leave the comfort of their ocean home and travel far from the sea. They have to reach a place where the ice is safe enough for them to raise their young. Their breeding ground can be up to

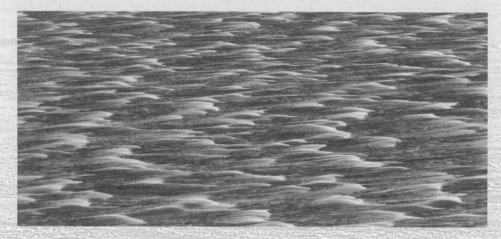

70 miles (113 km) away. But they can't fly there. Their wings make them excellent swimmers, but won't let them fly. So they must walk.

For the next nine months, they will march back and forth between the ocean and their breeding ground. Sometimes they will slide on their bellies as they go between the two, but for the most part, they will walk.

Their survival depends on this annual march. Much hardship lies ahead, and they will have many obstacles to conquer. And then there's the merciless cold, so cold that it blurs the vision and numbs all courage. But nothing is achieved without effort. The penguins must give themselves over to this struggle. They must surrender to this race against time. They will march until they are totally exhausted.

It is March. Summer is over. Another long polar winter is about to begin. The emperor penguins have been feeding in the relatively warm ocean waters for

# THE MARCH

three months. The ocean is already beginning to freeze over. The water at the ocean's edge will soon harden, forming a belt of thick ice. But the ice here won't be solid enough for them to mate and raise their chicks; a strong wind could destroy this ice in seconds. So it is high time for the penguins to go. They have to head south to their traditional breeding ground: the Adélie Coast. Time for one last swim, then the emperor penguins climb onto the ice. They line up in single file. With perfect discipline, the penguins fall in one behind the other. They puff out their white chests, displaying absolute calm. More and more penguins gather together. The great procession takes form. The last of the clan finally clambers up onto the ice. The march begins.

*Winter has already iced over the small islands of the Adélie Coast, freezing the sea spray whipped up by the latest storms. The emperor penguins will soon take over this landscape, where they will confront the harshest of winters.*

*The emperor penguins enter the city of ice, a jumble of hundreds of stranded icebergs.*
*The penguins march on, walking around these huge structures of ice that broke*
*off from the glaciers of the continent.*

The procession marches across the white desert in single file. Hundreds of penguins, all walking in perfect alignment: No penguin tries to pass another; each penguin walks in the tracks left by the penguin in front. The ice shines under the winter sun. The wind blows lightly, accompanying the sound of their claws against the ice. The emperor penguins leave the sea farther and farther behind. They walk slowly, able to take only short steps. But little by little, they make their way across the ice.

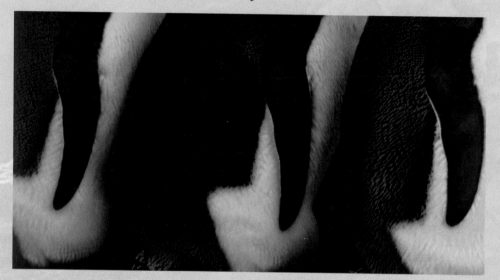

Night falls. They continue to march. The turbulent skies light their path, the ice glistening under the stars. To get to their breeding ground on the Adelie Coast, the emperor penguins will have to walk day and night.

Ahead is a skyline of ice. The emperor penguins come upon hundreds of stranded icebergs—a city of ice. This ephemeral city is enormous, majestic, intimidating. The penguins go around the massive blocks of ice, the first of many icebergs they will have to avoid along the way.

Their destination is always the same. Their path, however, is not. The ice never stops shifting and changing. New roadblocks appear every year. The latest hurdle: some hummocks, small pressure ridges of ice between two ice floes. With their beaks and claws, the emperor penguins tackle this new obstacle, probing the ridges until they eventually find a little strip of snow they can use as a ramp. They climb it, fall, get back up, fall again, crash into each other. Some of the penguins groan. Others, worn out, flop belly down on the ice. The cold stings, burns, freezes. The emperor penguins huddle together to keep warm. Each day, the temperature drops a little further. The sun is setting earlier. And the weather is becoming noticeably harsher—almost by the hour.

*After walking for nearly 70 miles(113 km), the penguins arrive in a protected enclave amid the tiny islands of Pointe-Géologie. Their long march is finally over.*

No one knows exactly how the emperor penguins find their way to their breeding ground. Perhaps they are assisted by the sun, or by the stars. Or maybe, having taken

# THE GATHERING

this march for thousands of generations, they are guided by some invisible compass within them. Their procession winds in a graceful series of curves across the white landscape. To keep warm, the emperor penguins hold their wings tightly against their bodies. Step by step, they near their goal. And after walking continuously for almost a week, they reach it. Finally: the Adélie Coast.

Safe harbor, at last. Other columns of penguins are arriving, too. Coming from every direction, the emperor penguins are all returning to the place where every one of them was born: the plain of ice between the small islands of Pointe-Géologie, an archipelago off the Adélie Coast. Over the next couple of weeks, 7,000 emperor penguins will gather here. Surrounded by ice cliffs, this oasis amid the white desert will provide them with a haven for their chicks.

*The ice between the small, rocky islands is reassuringly stable, firm enough for the penguins to raise their young here. But winter is arriving, with its storms and intense cold. The emperor penguins will have to struggle with the elements for nine months.*

In the shelter of this breeding ground, called an *oamok*—coined from the words oasis and hummock—the emperor penguins can protect their chicks from three dangers: predators, the winter winds, and cracks in the ice. The emperor penguins are now far from the water's edge, where most predators lurk. The large ice walls here offer some protection from the harshest winds. And the ice will stay solid until summer, which will keep the penguins chicks from accidentally falling into the freezing ocean.

Emperor penguins are monogamous. But most will usually pair up for only one mating season, the next year taking a new partner. Milling around the breeding ground, the emperor penguins begin their search for the mate they will have for the next nine months.

*Engaged in their courtship
dance, couples strike poses.
Through their mating songs,
they learn to identify one
another. After two penguins
choose each other as mates,
they adopt an ecstatic
position, becoming
oblivious to the noisy
crowd around them.*

*Silent and peaceful, newly formed couples isolate themselves, absorbed in their own little world. Through their courtship rituals, the couples develop very strong bonds and synchronize their movements, which coordinates their biological rhythms for the months to come.*

Emperor penguins don't have nests, which provide other species of birds with landmarks that help them locate and recognize their mate. Emperor penguins rely on sound, rather than sight, to recognize and find each other in the huge crowd of the breeding colony. Each penguin has a distinct vocal pattern. This vocal signature conveys the identity, sex, and breeding readiness of the penguin. When a penguin sings, the penguins nearby remain quiet. This protocol ensures that each penguin can take turns being heard. To find a partner, the male emperor penguins emit display calls. Pacing up and down the breeding ground, they stop, fold their necks down, and sing, trumpeting loudly. Then they raise their heads, growl, and continue walking if their call doesn't get a response. When a female penguin shows interest by trumpeting back, the two decide whether to pursue a courtship. The couples that pair off launch into a mating ritual. And as they sing and perform their mating displays, they learn to identify each other by their song.

Duos throughout the breeding ground are engaged in a magnificent ballet. They fold and unfold their upper bodies repeatedly, the bend of their torsos creating

# THE COURTSHIP

voluptuous curves. Mating songs ring out everywhere. But not all of the penguins are finding a mate. A violent quarrel breaks out among some females. They are barging in on a courtship, battling it out for the male. They shove, peck, thwack each other with their wings. Since the males endure the harshest conditions of the winter breeding season, and sometimes don't survive it, the colony has more females than males. When the courting period ends, the few females that didn't find a partner will have to head back out to sea.

The courtship displays will go on for another few weeks. Each stage of the penguins' mating cycle is a race against the clock—if any part of it isn't completed in time, the penguins won't have a chance to breed again until next year.

After two penguins have chosen each other as mates, they stop singing and moving, adopting an ecstatic position. Oblivious to the crowd around them, the pair remains motionless for several minutes. A strange muffled silence begins to reign over the colony. Motionless couples stand, here and there, at the edges of the rookery, while the females that are still unattached wander around the breeding ground, seeking the last remaining available males. Tender duos, bright white spots under a sky drenched with stars, walk in tandem, waddling from side to side in concert. Looking blissful, no longer aware of the blizzard or the cold, they stand close to one other, eyes closed.

In the dense, noisy crowd of the colony, it is easier for the emperor penguins to locate each other by sound rather by sight. As soon as two penguins have formed a couple, they stay by each other's side—no question of taking the risk of losing the mate they have just chosen.

*The females' eagerness to find a mate sometimes turns combative.*
*Hitting and pecking, the females that haven't found a partner fight*
*for every remaining available male.*

The couplings bear fruit in May and early June. Each female penguin will lay a single egg. The female lays her egg standing up, a particularly risky maneuver.

# THE EGG

The egg—the treasure—must under no circumstances roll on the ice. Within minutes, it would freeze. The female penguin watches tensely. If she fails, this is what awaits her egg: a carpet of ice, hard, bright, sharp, like a diamond. Exposure to the freezing air for more than an instant would destroy this magic embryo, this little being still imprisoned in its shell. She has already prepared a shelter for her egg in a fold in her stomach. A few minutes more, and here comes the egg. Cold outside, warm inside. She slides the egg onto her feet and immediately hides it in her incubating pouch. The next day, the penguin couple undertakes the difficult work of transferring the egg.

The egg needs to be incubated. But the female, exhausted from producing and laying the egg, has lost almost a third of her body weight. She must quickly return to the sea, or she will die. And while she feeds in the ocean, she will not only be rebuilding her strength, she will be gathering food to bring back for her chick. The couple will split the tasks of tending to their egg: She will go in search of food, and he will have the enormous responsibility of keeping this little bit of life safe and warm. The male will hold the egg on top of his claws, under a flap of skin on his stomach, for more than two long months of deprivation and struggle to preserve life.

Following a very precise ritual, the female must transfer her precious cargo to the male. The female penguin backs up and lets the egg roll on the ice. The male, with the help of his beak, immediately pushes the egg onto his feet. The two partners have to act in perfect synchronization. One step too many, one hurried movement, the slightest error, and the egg will freeze.

*Weakened by egg laying, the females leave the colony and begin the long, solitary march back to the sea, where they will feed. Racing to arrive there before the weather turns harsher, they choose the shortest route.*

*Captivating jewels decorate the ice in winter. Antarctica is strewn with icebergs, which broke off from glaciers. The ice of some icebergs, subjected to tremendous pressure when the glaciers formed, is so dense that no air has been able to penetrate it. As a result, these icebergs display every shade of blue.*

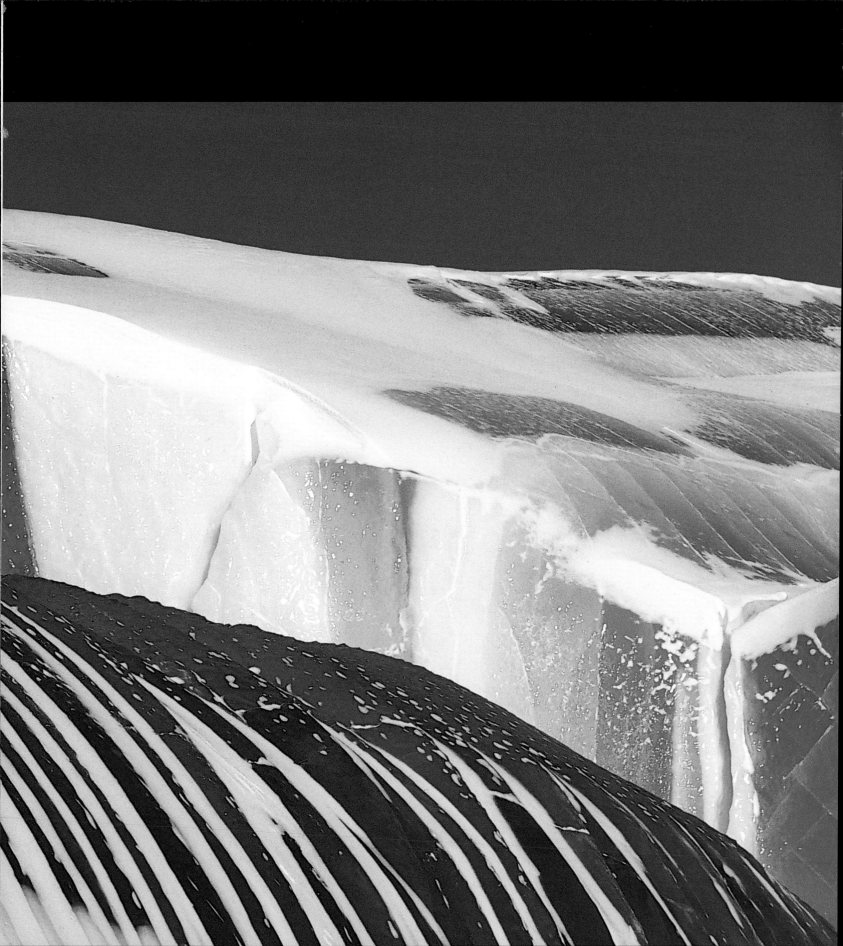

Before they transfer the egg, the couple rehearses each step of the procedure. Even with practice, mishaps occur. The couples who lose their egg will return to the sea, their partnership now over. At the edge of the colony, dozens of eggs, destroyed and frozen, testify to the difficulty of moving the egg between the two partners.

The days are getting shorter now. The females must return to the sea before the worst of the winter arrives. This march will be far harder than the first march. It is much colder, and the females haven't eaten in months and are depleted from bringing forth their egg. But they must hurry back. Time to part. One last, heartrending song: The partners sing to each other, a final song before separating to guarantee that they will recognize each other's voice when she returns to the breeding ground two months from now. The females, alone or in small groups, begin the trek back. Soon, they disappear into the horizon.

*A break in the ice reveals a polynya, a stretch of open sea. Rivers cut long grooves between the slabs of sea ice.*
*All of these openings to the water are a windfall for the females on their route back, providing food.*

For the males, a true ordeal now begins. They gather together. Walking on their heels, their egg wedged on their feet, they huddle into a group. This formation, called a turtle, helps them withstand the cold. Standing head to head, their eggs carefully protected under their stomachs, they settle in for their long wait.

A winter storm arrives. The temperature is now 60 degrees below zero(-15.5°C), even colder if the wind is taken into account. The wind, unleashed, is blowing 100 miles(161 km) per hour. The catabic winds, stronger in Antarctica than anywhere else, are rolling in from the South Pole. These cold winds, growing fiercer as they make their way across the continent, are at their most powerful when they get to the coast. The colony at Pointe-Géologie feels their full blast. Snow flies everywhere. Soon, a whiteout occurs. The cluster of penguins can't be seen even from 15 feet(4.6 m) away.

The couples' separation will last for more than two months. For the males, the long wait has just begun. United in the fight against winter, they huddle together in a turtle formation, making a rampart of their bodies against the blizzards and the extreme cold.

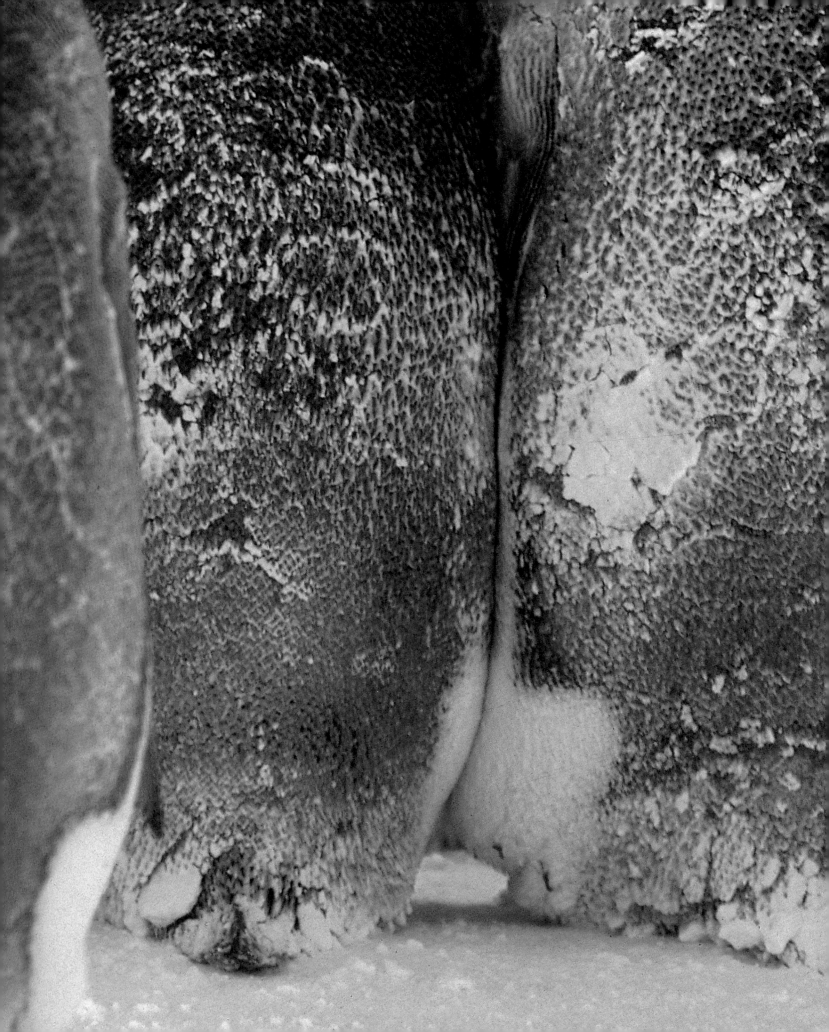

The nights are endless. Winter is now at its harshest. Daylight lasts only a couple of hours. The emperor penguins are tired, fragile. Every once in a while, the group, which is usually calm, becomes agitated and disorganized. One of the males stumbles. All of the heads in the huddle go up at the same time. The disruption causes some of the penguins to try to move out of the turtle. Total panic ensues, but it soon subsides. The turtle regroups several yards away, regaining its orderly, geometric pattern. In the old location, however, there are remnants of the upheaval: The ice is littered with eggshells, cracked open by the freezing cold. The days pass. The turtle, moving according to the direction of the wind, relocates a short distance away, leaving the soiled traces of its passage.

The males endure up to four months
of fasting during the harshest part
of winter. Their fate hangs on the
successful return of the females.
Patiently making short trips, they
walk with their eggs carefully
wedged under their incubating
pouches, moving as a group to
a clean new area of the ice.

Seventy miles (113 km) away, the females, exhausted as well, are caught in a violent nighttime storm. Despite the blizzard, they have to keep walking. They have marched back to where they started three months ago. But new ice has formed along the water's edge. The sea is now much farther away. The next day, they become aware that they are getting closer to the sea. In the distance, a dark shape stands out—it's the reflection of a large body of water in the middle of the ice: a polynya. The females arrive at its edge. Starved, they dive in, eager to feed on fish.

In the deepest part of the polynya, they come across some whales and Weddell seals, also on the hunt for food. The penguins, awkward on land, are once again in their element. Able to hold their breath for more than 15 minutes, they can dive as far down as 1,480 feet (450 m). For the next few weeks, they will feed on fish, krill, and squid, gathering nourishment for themselves and their chicks. They forage deep in the water and also along the undersurface of the ice. For now, the female penguins must rebuild their reserves of fat. Later, when they are strong again, they will march once more, returning inland to go to the aid of their partners and chicks.

After days with almost no light, the sun now begins to return to Antarctica—not for long, but enough to signal that winter is starting to recede. The male

# THE BIRTH

penguins are hungry and cold, but they have to keep moving to stay alive. Some of the males, mostly the older ones, haven't survived the long, hard months of fasting and incubation. In mid-July, some muffled chirps come from beneath an incubating pouch. A little head steals out, begging for food. Then it quickly disappears, hiding under the father's downy feathers. In the next few days, more and more tiny heads dart up around the colony, all peeping for food from between their fathers' feet.

After hatching in July, the chicks peek out
to demand their first beakfuls of food. A
chick will stay in the father's incubating
pouch only a few days at most. Mothers
return from the sea to the breeding ground
soon after the chicks hatch, and with the
fathers they alternate feeding and brooding
responsibilities over the next few months.

The father's incubating pouch keeps the chick warm, protected from the wind and the cold. But the chick is hungry. And the father is, too. If the mother doesn't return from the sea soon, the father will have to leave the chick behind and go back to the ocean to feed. Any longer without food, and he won't have enough strength to survive the trip.

To preserve the life of his offspring, the male performs an amazing feat: He regurgitates a milky substance from a crease in his throat, stored there especially for the chick. This small, nutritious meal will buy some time, keeping the chick alive for a few days.

*A chick stays in the shelter of its parents' incubating pouches for 45–50 days. Fed by their parents, the chicks grow quickly, constantly clamoring for more food. Soon the chicks will go out to explore their surroundings.*

On the horizon, an army is on the march: The females are back. Heads lift up from the turtle formation. The males start to sing, first a soft murmur, then a huge clamor as the females get nearer. Each of the females seeks her mate, trumpeting loudly. Her mate, recognizing her song, trumpets back. Despite the noise all around them, they find each other. Their calls rouse their chick, who peeps, hungry. Wasting no time, the female feeds her chick. Then, in a transaction as delicate as when they transferred the egg, the parents move the chick from the father's feet to the mother's. Any mistake here, and the chick could freeze. The father has to head out to sea as quickly as possible, but first, he sings. The chick sings back. Now, the two will be able to recognize one another when the father comes back.

The males, famished, now begin their long walk back to the sea—70 miles(113 km) on an empty stomach. They haven't eaten for months. Sometimes, when the wind brought snow, they had a chance to have something to drink. But they haven't had food of any kind since leaving the sea, losing as much as half their weight. Weary and drooping, the males head for the ocean, walking in the half-light.

Some will succumb to a blizzard or to exhaustion. Others will reach the sea but not survive the dangers there, such as leopard seals. These predators also snatched away some of the mothers. If one of the parents dies, the chick will die, too: Penguin chicks can be fed only by their own parents, and since the parents have to spell each other, if one of them is late or doesn't come back, the chick won't be able to live. Weak, the males walk back to the sea, alone or in small, silent groups. Their hard march back may account for why the colony has fewer males than females.

*After their winter vigil over the egg has ended, the males head straight to the sea. Exhausted, they have to put all of their remaining strength into making the 70-mile(113 km) trip. Many won't survive it.*

Under a deep blue sky and pearly clouds, the colony bursts with life. The chicks are growing stronger. Some have ventured out from the incubating

# THE FIRST STEPS

pouch. Already the chicks want to break out of the safe shell of their parents' care and start discovering the world. The snow, the icebergs, the ice — they don't know anything about these dangers yet. But the chicks are inquisitive and daring, eager to explore. Some of the mothers, impatient to return to the sea and weary of waiting for their mate's return, temporarily abandon their chicks. These chicks huddle together to keep warm, a baby version of the turtle formation. Their little group creates the hub of what will become the nursery.

*The chicks are now independent of their parents, but are anxious about the separation. Their parents make frequent trips back and forth to the sea. The chicks watch for their return.*

*During their parents'
prolonged absences, the
chicks are on their own.
They protect themselves
from the cold by huddling
together in the nursery.*

The rest of the chicks in the colony have joined the little huddle. The chicks no longer need the protection of being carried on their parents' feet. Together, the gathered chicks keep each other warm. And having each other's company soothes them during their parents' absences. Several weeks after the chicks' birth, the fathers return. It now becomes the mothers' turn to go to the ocean to get food. If the mother has already left, the father looks for his chick in the nursery.

The father and chick recognize each other from their songs. Among all the downy little chicks in the nursery, some clamoring for food, beaks wide open, a father is able to tell which is his chick. And he won't feed any chick but his. The parents now begin to travel back and forth to the sea more frequently—a necessity to keep the hungry chicks fed. During their incessant comings and goings, the parents are stocking enough food to give the chicks their fill.

A thick layer of snow has fallen on the nursery. The squall is sweeping the snow over the chicks, blowing it over them in great gusts. Far from the safety of the incubating pouch, the chicks are cold. But they can't send a call for help to the adult penguins. Huddled together, the adults have drifted too far away, pushed some distance from the nursery by the blizzard. The rampart of their dark backs is barely discernible during the brief periods of visibility. The chicks, panicked, break apart from their turtle formation and scatter in all directions. Chaos. Some of the chicks manage to reach the adults' huddle. But other chicks are wandering around, shivering, peeping in distress. Where can they turn for help? Winter—their very first winter—is going to get the better of them.

The next morning, under a bright, indifferent sun, dozens of little bodies lie scattered at the edge of the colony. A few yards away, a large crack runs through the ice, fractured by the storm. The chicks find themselves separated from their parents, who call out to them and encourage them to confront this new danger. But soon, the jaws of ice close up again.

The movements of the sea have opened up the ice. A crack in the ice threatens the colony.
The parents sing to their frightened chicks to encourage them to cross these precipices.
This is how their youngsters learn about life on the ice.

The parents of the chicks lost in the storm search for their young amid the bodies on the ice. The loss is unbearable. Every year, some bereft mother will respond to her anguish by trying to steal another mother's chick.

Although she has lost her own chick, she continues to have the instinct to feed and nurture a chick. One bereft mother is on the lookout for a lone chick. She sees a chick standing apart from his mother. She tries to take him away. But several penguins come to stop the abduction. The tugging and shoving lasts several minutes. And soon the chick is back in his mother's care.

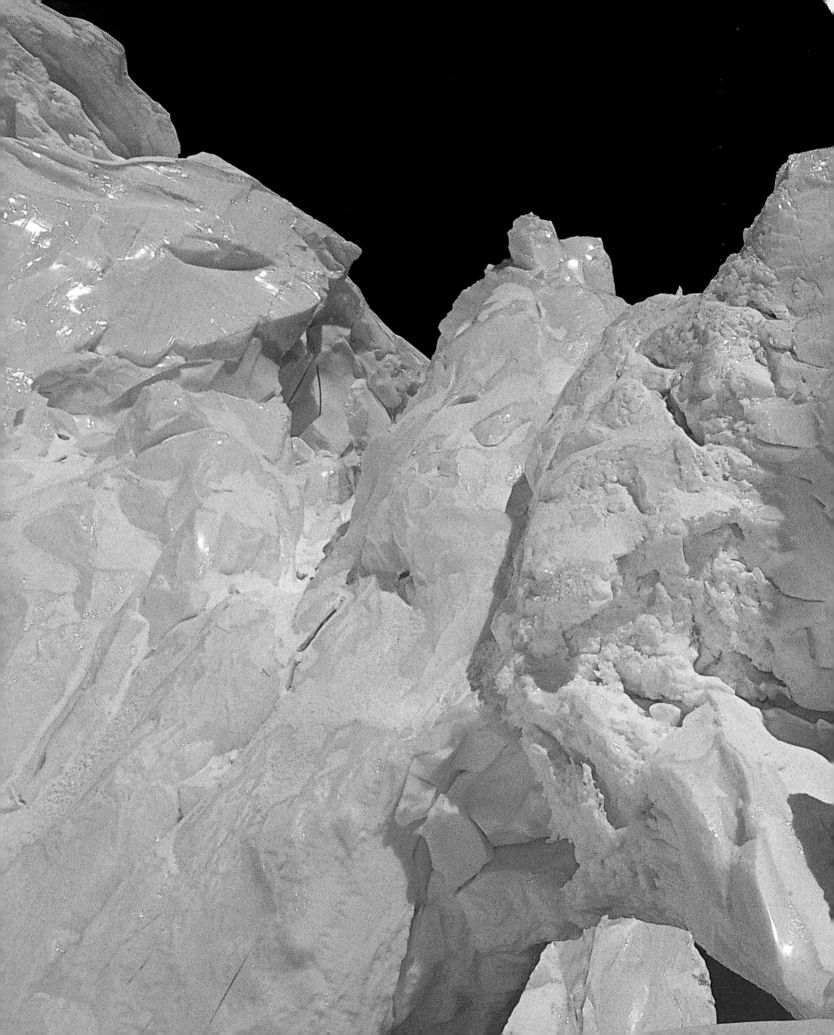

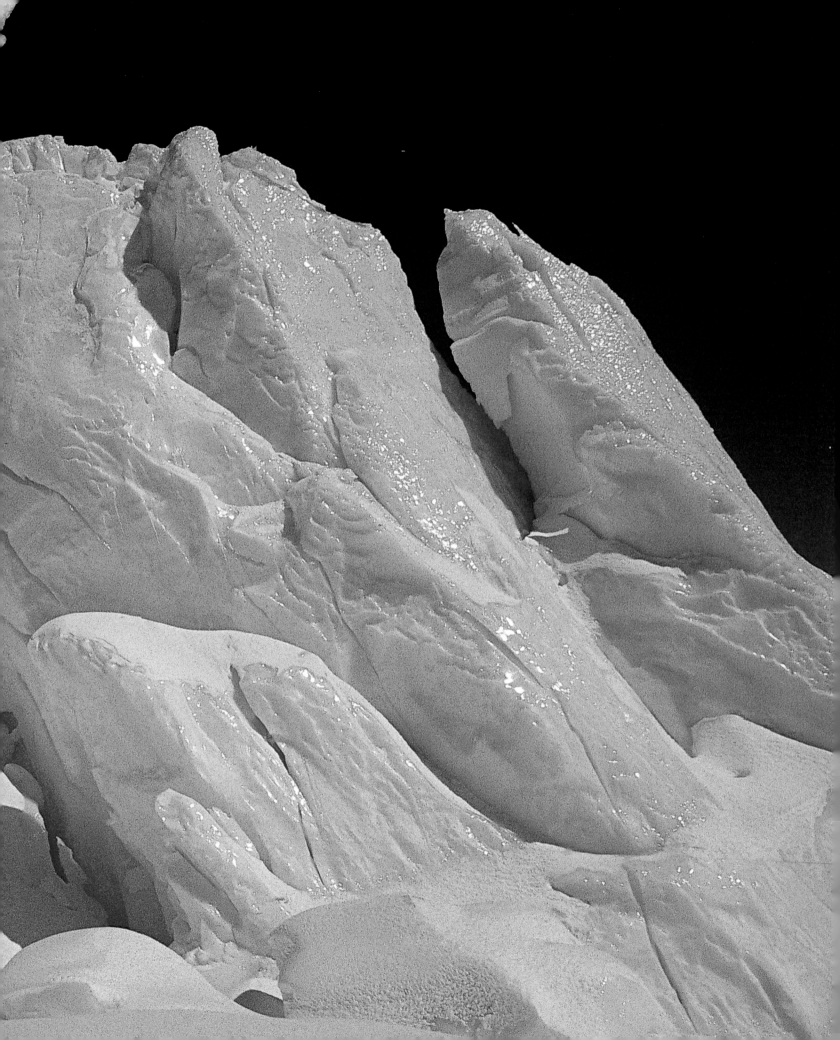

The weather finally warms up. Large cracks furrow the ice. The slabs of ice rub and grate against each other, producing strange whistling sounds. The clear sky

# THE SPRING

presages milder days ahead. In September, spring offers its riches: It is no longer necessary to travel long distances to the sea for food; the sea is now nearer. Day by day, the chicks are growing, becoming more and more independent. Soon they will be as big as their parents! They are thriving in the nursery and going farther afield, scattering out on their own to discover life.

*With the melting of the ice, the time to leave approaches. As the chicks molt, their parents will gradually stop feeding them. The chicks are now five months old; soon they will be completely on their own.*

But spring also brings its share of risks. An immense bird arrives, gliding into the colony, and lands just a few dozen yards from the nursery: It's the giant petrel, the first visitor of the spring. The vulture of the Antarctic circles around the huddle of chicks. A lone chick toddles within a few yards of the petrel. With one hard blow of his wing, the petrel knocks the chick down. And with his powerful beak, he twists the chick's neck. The petrel shakes the chick frenziedly, then gulps the chick down. The adult penguins stand by, doing nothing. They understand: This first murder sounds a warning. The petrel, the bloodthirsty tenant of a piece of the sea, has taken up residence for a few weeks.

*In September, with the arrival of spring, the giant petrel arrives suddenly from the sea and prowls on the edges of the colony. The petrel will claim his share of isolated chicks. Unable to eat the numerous frozen corpses, the vulture attacks the living.*

It is now October. Spring is asserting itself across Antarctica, to the great delight of the emperor penguins. Under an increasingly hot sun, the ice melts bit by bit, releasing the icebergs, which gleam with light. The rocks of the small islands now are mostly bare of snow. In only two more months, the young penguins will be able to dive into the water to go live in the sea for four years—time for them to reach their full size. The ice has completely disappeared, opening the way to the ocean and unveiling a sky-blue sea, buffeted by the foam of gushing waves. Both the little and the big emperor penguins have now found their element again: the water.

*Some blocks of ice start to give way under the chicks' feet. Soon, all of the ice will disappear, forcing the chicks to take to the water.*

The adults show the way. They dive into the water in groups, romping with delight in the clear water. Finally, to live free and in the sea! Three months of

# THE PLUNGE

summer, three months of freedom. The young penguins will remain here for a few more weeks, growing stronger. It is now December, and they are ready to leave the place where they were born. They take the plunge, and go home for the first time. For the next four years, they will live at sea, and then they, too, will continue the tradition. They will climb out of the water and they will march, just as emperor penguins have done for centuries.

*The ice explodes, a disaster. But by shifting, the ice opens the way for the chicks to start their journey toward the unknown: the long voyage to the open sea, where they will stay for four years.*

*Anxious, but making their great departure, the chicks dive into the water.*

No one knows where the emperor penguins will go for the next three months. North? South? Now, that's another story. But one thing is sure: Next March, they will be here.

To tell the story of this extraordinary bird, we have to cross two hemispheres and isolate ourselves for a year in a place with the worst winters on the planet. In Antarctica, in the kingdom of ice, the emperor penguins take over the ice for nine months to give life. To film their incredible effervescence, we let ourselves be shut in by the Antarctic winter, and are witnesses, day after day, to this astonishing animal's permanent struggle with the elements. We live this adventure in the archipelago of Pointe-Géologie, where the blizzards and the polar night reign, and we live it with the emperor penguins, who accepted us and guided us in our progression through winter.

Though shooting with a small crew, we profit from having a scientific base nearby. We spend the winter with about 30 people at the French polar station Dumont d'Urville, on a tiny island at the foot of Antarctica, a few hundred yards from the south magnetic pole. This station is isolated from the world during the winter months, when the extreme cold ices over the area. The frozen sea then surrounds the continent with a belt of ice up to 125 miles (200 km) wide.

A colony of 7,000 penguins has an annual rendezvous with winter at the foot of the Astrolabe glacier, coming there every year at the end of March. We are on their trail. As soon as they arrive, everything changes. The silence of Antarctica is disrupted by the birdsongs of thousands of

penguins. The emperor penguins lead the dance and impose their own rhythms. Our days and nights become organized according to their behaviors.

Our daily appointment with the penguins becomes a ritual. Each morning we join them to spend the day in the colony. It is April; it's four degrees below zero (-20°C). There is a haunting cacophony. The emperor penguins are meeting, courting loudly, forming couples. We are the privileged witnesses of their rendezvous of love. To our great surprise, the penguins come toward us and surround us. Their innocence enchants us: There are still living beings on this Earth who aren't alarmed at the presence of humans. It even seems as though some of them are trying to charm us. We are finally adopted by a pair of penguins, whom we learn to recognize in the coming days. Becoming familiar with our orange outfits and our equipment, they will accompany us for the rest of the year.

*Water is everywhere, frozen by winter into dazzling forms. The light plays on the surface of the icebergs and transforms the ice into crystal. The frost creates the most extravagant forms. This spectacle delights us,*

Thanks to the emperor penguins, we find our way in this kingdom of ice. We follow them on the first few miles of their journey between the sea and their colony. They guide us through the labyrinth of icebergs. We lose our way, amazed by the confusion. Each day we are fascinated by these natural constructions with monumental proportions. We approach some cathedrals of ice that exist for just the length of a winter; inside, we penetrate the blue heart of immense icebergs.

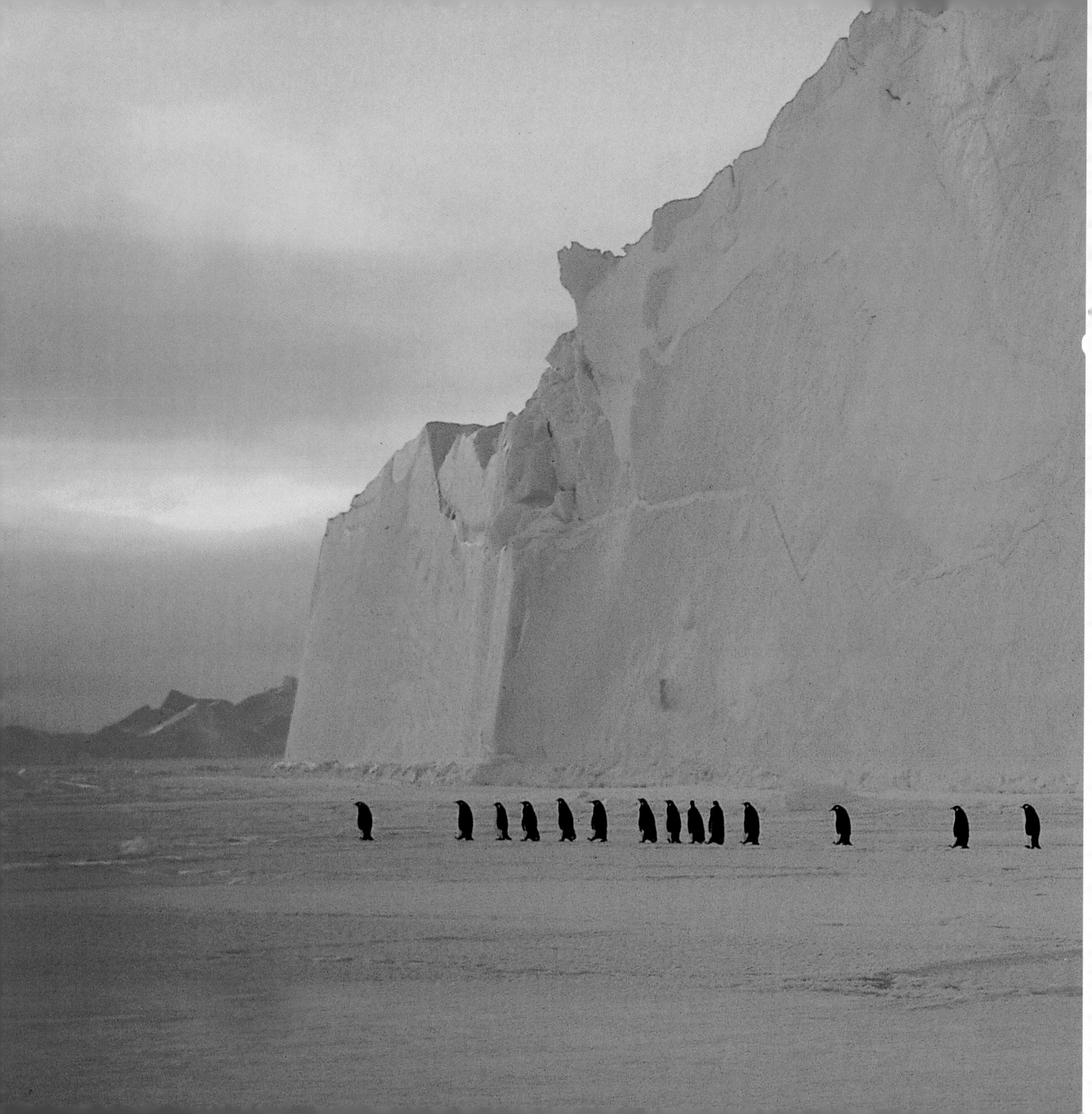

We come to know the Antarctic. Wanting to film in all kinds of weather, we are forced to adapt to the long polar night, as well as to the terrible unpredictability of the catabatic wind. This wind pulverizes the snow and makes the colony disappear under a cover of white. During these interminable blizzards, we have a great deal of trouble filming our scenes. At 40 below zero, each gesture costs us, and we can appreciate how much the penguins have to endure. In winter, the storms intensify, sometimes settling in for several days. We are frequently roughed up by storms while we haul our 132-pound (60 kg) *pulkas* across the ice. But the play of light and the surreal colors of the landscape continually attract us.

We quickly become the guests of the ice and the penguins, adopting the habits of the inhabitants of the ice. Preparing lunch on the ice becomes a disconcerting ritual. In this cold, the simplest actions are transformed into true ordeals. Feeding ourselves

becomes a chore. But how can we complain about this food that we can never manage to heat, when the penguins right next to us are on their third month without food?

The emperor penguins know us and tolerate our presence quite calmly. And we also can't resist our desire to film the colony from the sky, using a telescopic crane. We fly over the 7,000 penguins huddled against the cold, and we can zoom in on the rookery in the search for the first chicks. At the end of the long polar night, these new filming possibilities fill us with enthusiasm. In the midst of winter, Antarctica appears to us as a region where light is rare, beautiful, and precious.

Finally, it is the penguins themselves who give us the signal to leave. When the ice recedes, the *Astrolabe* returns. Its stern breaks up the ice, opening the path for our return home. Despite the 13 months we've spent on the ice, it seems as if we have just skimmed the surface of Antarctica. The boat takes us back to the noisy world where time is measured differently than by following the arc of the sun. We don't yet know how hard it will be to return and become citizens among men, after these long months when we have lived as subjects of the emperor.

# MARCH OF THE PENGUINS

**Published by the National Geographic Society**

| | |
|---|---|
| John M. Fahey, Jr. | President & Chief Executive Officer |
| Gilbert M. Grosvenor | Chairman of the Board |
| Nina D. Hoffman | Executive Vice President President, Books and School Publishing |

Prepared by the Book Division

| | |
|---|---|
| Kevin Mulroy | Senior Vice President & Publisher |
| Kristin Hanneman | Illustrations Director |
| Marianne R. Koszorus | Design Director |
| Rebecca Hinds | Managing Editor |

Staff for this Book

| | |
|---|---|
| Lisa Thomas | Editor |
| Cameron Zotter | Production Artist |
| Emily McCarthy | Researcher |
| Michael Greninger | Researcher |
| Judy Klein | Contributing Editor |
| Gary Colbert | Production Director |

Manufacturing and Quality Control

| | |
|---|---|
| Christopher A. Liedel | Chief Financial Officer |
| Phillip L. Schlosser | Managing Director |
| John T. Dunn | Technical Director |

Founded in 1888, the National Geographic Society is one of the largest nonprofit scientific and educational organizations in the world. It reaches more than 285 million people worldwide each month through its official journal, National Geographic, and its four other magazines; the National Geographic Channel; television documentaries; radio programs; films; books; videos and DVDs; maps; and interactive media. National Geographic has funded more than 8,000 scientific research projects and supports an education program combating geographic illiteracy.

For more information, please call 1-800-NGS LINE (647-5463) or write to the following address:

National Geographic Society
1145 17th Street N.W.
Washington, D.C. 20036-4688 U.S.A.

Log on to nationalgeographic.com; AOL Keyword: NatGeo.